# 百年掠影
## ——山东近代建筑集萃
COLLECTION OF MODERN ARCHITECHTURES IN SHANDONG

■ 李华文／主编
■ 周玉山·马月华／副主编

济南出版社

# 《百年掠影——山东近代建筑集萃》编委会

主　编　李华文

副主编　周玉山　马月华

编　委　李华文　李西宁　王寿宴　王丕琢
　　　　梁泽庆　周玉山　马月华　张汝锋
　　　　陈　宾　王亮朝　蔡小晶　于明良
　　　　杨泽宇　刘元安　张泉刚　王海明

摄　影（排名不分先后）

| 鲍颖超 | 陈海军 | 李华俊 | 刘　刚 | 刘　杰 | 刘子维 | 吕维君 | 王　伟 | 习娟娟 | 张　卡 |
| 张　磊 | 赵清彬 | 赵文鑫 | 韩春山 | 刘吉亮 | 吕玉珍 | 孙崇伦 | 杨　霞 | 蔡　辉 | 陈光东 |
| 陈宏伟 | 陈开元 | 陈　鑫 | 陈杏云 | 陈　燕 | 程普麟 | 程永林 | 程元林 | 崔文斌 | 单　勇 |
| 邓树江 | 刁伟明 | 董　佩 | 范一飞 | 方志广 | 付庆军 | 高　超 | 高龙雨 | 谷忠民 | 顾　青 |
| 关　川 | 郭　旗 | 侯晓燕 | 胡光艳 | 胡　明 | 黄　平 | 蒋士杰 | 金　磊 | 荆　强 | 孔　军 |
| 匡　松 | 李承东 | 李　刚 | 李文华 | 李家荣 | 李建辉 | 李金刚 | 李　康 | 李　乐 | 李　力 |
| 李连昇 | 李　亮 | 李延伟 | 李玉鸿 | 刘春明 | 刘　军 | 刘　琴 | 刘　伟 | 刘玉华 | 陆晓林 |
| 马广德 | 马卫国 | 孟现雷 | 孟祥晶 | 饶　琦 | 任　军 | 孙济生 | 沈永岭 | 宋洪晓 | 宿　燕 |
| 孙　华 | 唐卫平 | 王春燕 | 王海军 | 王洪章 | 王慧燕 | 王　军 | 王　敏 | 王丕琢 | 王　平 |
| 王　琴 | 王晓菲 | 王言法 | 王一飞 | 王玉祯 | 王振萍 | 王志斌 | 王子辰 | 熊幼玟 | 徐金利 |
| 徐　艺 | 闫　济 | 杨明鑫 | 杨万智 | 杨宇晗 | 于　骞 | 于　青 | 于晓明 | 袁平建 | 袁雅民 |
| 翟华清 | 张宏毅 | 张　火 | 张晓晨 | 张旭辰 | 张钊疆 | 张智平 | 赵厚才 | 赵　静 | 朱　军 |
| 朱明兰 | 朱伟杰 | 李秀平 | 惠荻林 | 孙　广 | 丛吉东 | 杜成立 | 付仁峰 | 卢翠华 | 马卫东 |
| 赵　翔 | 赵新宏 | 关义华 | 薛清峰 | 刘国强 | 刘建国 | 杨保胜 | 樊　凯 | 耿志刚 | 纪才溪 |
| 张道勇 | 陈昌礼 | 陈　翔 | 刘　涛 | 刘　璇 | 刘兆春 | 苗　莉 | 苏　健 | 孙文丽 | 王新正 |
| 王旭生 | 王昭脉 | 薛修利 | 杨光远 | 张建华 | 张宗贤 | 赵国良 | 赵小兵 | 赵紫萍 | 郭　梁 |
| 薛信喜 | 李　冰 | 乔宏伟 | 王传慧 | 丛建萍 | 董晓黎 | 李段华 | 刘巍峰 | 刘向城 | 刘雅菁 |
| 曲　明 | 于恒江 | 赵丽金 | 李琳琳 | 薄淇元 | 冯建国 | 高子会 | 郝文强 | 焦文华 | 金志远 |
| 刘光来 | 刘　媛 | 刘　忠 | 路仁玲 | 石　磊 | 宋长春 | 谭兴华 | 王伯林 | 王应刚 | 肖海军 |
| 谢啸林 | 杨向东 | 杨增华 | 衣汉叶 | 尹衍河 | 赵江涛 | 常　颢 | 迟少红 | 李爱山 | 林裕超 |
| 刘福轩 | 刘国哲 | 柳艳梅 | 苗　伟 | 宋石刚 | 孙国明 | 孙　强 | 王俊杰 | 王奎乐 | 张卫国 |
| 李希元 | 陈允沛 | 洪晓东 | 贾洪章 | 贾宜锦 | 梁克霞 | 武靖力 | 夏小敏 | 张　伟 | 凌小凡 |
| 鲍媛媛 | 李　波 | 孙　磊 | 吴鹏程 | 吴　伟 | 严承银 | 谭　国 | 赵紫萍 | 张　军 | 鲁守钦 |
| 赵兴普 | 费圣义 | 王汉东 | 洪亚彬 | 胡德定 | 武明理 | 房晨生 | 孙　岩 | 王海明 | 张　伟 |
| 赵君武 | 钟志忠 | 董丽娜 | 张泉刚 | | | | | | |

# 定格近代建筑 见证百年沧桑（代序言）

山东近代建筑是指于一八四〇～一九四九年期间山东境内建成的各类建筑。建筑风格涵盖西式建筑、日式建筑、中西结合建筑、传统中式建筑等，建筑类型包含传统民居、戏院、图书馆、博物馆、学校、医院、教堂、领事馆、银行、邮局、营房、钟楼、车站、体育场以及工业性建筑、纪念性建筑等。

山东近代建筑资源极为丰富，其特点是分布范围广、数量众多、风格流派多样。山东近代建筑根植于中国传统的建筑文化和建筑艺术，同时又吸纳了西方建筑流派和艺术风格的诸多元素，呈现出建筑风格中西融合、建筑艺术精彩纷呈、建筑形式多样化的基本特征。

建筑是凝固的历史。山东近代建筑铭刻着近代山东历史变迁和社会发展的沧桑印记，承载着极为丰富的历史文化内涵，是一笔宝贵的文化遗产。近代建筑物在我省分布广泛，数量众多，通过精心组织、广泛发动，全省十七市均有作品参赛。参赛作品既有列入各级文物保护单位的知名建筑，也有新发现的近代建筑物，基本反映了山东各地近代建筑的全貌。二是建筑类型齐全。参赛作品既反映了代表传统建筑文化的楼堂庙宇、宅院民居，也有列强殖民入侵时期所建的教堂、领事馆、车站、兵营、学校、医院、银行、邮局、商店、工业建筑等，以及民国以后陆续建设的中西合风格的城市建筑。三是作品来源多样。既有参赛作者近期拍摄的作品，也有摄影师们多年以前的压箧旧作，还有具有历史价值的黑白老照片。四是大赛参与面广。参赛作者既有专业摄影工作者，也有热心的民间业余摄影爱好者。大赛组委会还多次组织专业摄影工作者到有关市县进行创作采风，有效地保证了参赛作品的质量。

从二〇一二年十月开始的山东近代建筑艺术摄影大赛，是由山东省艺术摄影学会、山东省艺术馆主办，山东艺术摄影网、济南九创公司、济南市摄影家协会承办的。经过近一年的努力，先后收到各地参赛作品近万幅，涉及山东各地近代建筑物近千处。

山东近代建筑艺术摄影大赛，是我省摄影界首次就这一题材组织举办的摄影盛事，其宗旨是『定格近代建筑历史，传承百年建筑文化』，目的是运用摄影这一独特的艺术形式，再现近代建筑的魅力，留存近代建筑的精彩身姿。大赛得到了社会各界的认同和支持，得到了摄影家和广大摄影爱好者的积极响应和热情参与。此次活动特点鲜明：一是作品涉及范围广。

此次出版的画册，便是从山东近代建筑艺术摄影大赛精选优秀作品的一个结集，也是我们探寻和研究山东近代建筑的一项初步成果。时间仓促，缺憾之处不可避免，但我们仍然冀望这个结集能够引导人们认知近代建筑，走近近代建筑，感知这些珍贵文化遗产带给人们的美丽和沧桑。

编 者

二〇一三年十二月

# 目录

## 纪念性建筑

济南·解放阁 \ 001
济南·五三惨案蔡公时殉难地 \ 002
济南·五三惨案纪念碑 \ 003
济南·辛亥革命烈士陵园 \ 004
济南·辛稼轩纪念祠 \ 005
济南·奎虚书藏 \ 006
济南·浙闻会馆 \ 006
济南·虎豹川修路碑亭 \ 007
济南·张氏祠堂 \ 008
济南·广智院 \ 009
泰安·冯玉祥读书楼 \ 010

## 政府·机构

临沂·山东省政府旧址 \ 019
泰安·中共东平县工委诞生地 \ 018
潍坊·潍坊特别市委旧址 \ 018
济南·欧人监狱旧址 \ 017
青岛·德国警察署旧址 \ 016
青岛·德国总督府旧址 \ 015
青岛·胶州帝国法院旧址 \ 014
济南·中共济南乡师支部诞生地 \ 014
济南·中共平阴县委旧址 \ 013
济南·中共长清县委临时党支部成立旧址 \ 012
济南·中共山东省委秘书处旧址 \ 011

## 军事设施

青岛·德式官邸旧址 \ 022
青岛·德国第二海军营部大楼旧址 \ 021
济南·济南战役山东兵团指挥所旧址 \ 020
莱芜·莱芜战役指挥所遗址 \ 029
临沂·八路军一一五师司令部旧址 \ 029
临沂·新四军军部旧址 \ 028
临沂·八路军一一五师司令部旧址 \ 027
威海·军官食馆处旧址 \ 026
威海·华勇营大楼旧址 \ 025
威海·东泓炮台 \ 025
威海·北洋海军提督署 \ 024
烟台·西炮台 \ 023

## 领事馆

烟台·英国领事馆附属建筑旧址 \ 039
烟台·日本领事馆旧址 \ 038
烟台·美国领事馆领事官邸旧址 \ 038
烟台·挪威领事馆旧址 \ 037
烟台·俄国领事馆旧址 \ 036
烟台·丹麦领事馆旧址 \ 035
青岛·湛山三路288号建筑 \ 034
青岛·美国领事馆旧址 \ 034
济南·英国领事馆旧址 \ 033
济南·济南日本总领事馆旧址 \ 032
济南·德国领事馆旧址 \ 031
济南·美国领事馆旧址 \ 030

## 商行·商埠

烟台·五洲大药房旧址 \ 048
烟台·万国理发馆旧址 \ 047
烟台·顺昌商行旧址 \ 047
烟台·庆昌五金行旧址 \ 046
烟台·法商永兴洋行旧址 \ 046
烟台·瑞蚨祥旧址 \ 045
淄博·仁德茶庄旧址 \ 044
青岛·湛山三路287号建筑 \ 043
青岛·礼和商业大楼旧址 \ 042
济南·原皇宫照相馆 \ 042
济南·瑞蚨祥鸿记旧址 \ 041
济南·宏济堂西记 \ 041
济南·复成信东记、西记 \ 040

## 车站

济南·北关车站旧址 \ 078
济南·济南火车站 \ 077

## 学校·医院

临沂·美国教会医院旧址 \ 076
滨州·英国教会医院旧址 \ 075
威海·威海卫学校医院旧址 \ 074
潍坊·乐道院医院十字楼 \ 073
烟台·益文商业专科学校旧址 \ 072
烟台·崇正中学旧址 \ 071
烟台·光被中学旧址 \ 070
烟台·市立烟台医院旧址 \ 072
淄博·同仁会济南医院旧址 \ 069
济南·山东省红卍字会施诊所 \ 068
济南·正谊中学 \ 067
济南·懿范女子中学 \ 066
济南·原齐鲁大学近现代建筑群 \ 065
济南·原济南市第一中学教学楼 \ 064

## 银行·邮局

烟台·烟台邮局旧址 \ 063
烟台·交通银行烟台支行旧址 \ 062
烟台·汇丰银行烟台分行旧址 \ 061
烟台·德国邮局旧址 \ 060
青岛·青岛湛山三路293号建筑 \ 059
青岛·湛山三路290号建筑 \ 058
青岛·交通银行青岛分行旧址 \ 057
青岛·帝国邮政局旧址 \ 056
青岛·德华银行旧址 \ 055
青岛·大陆银行旧址 \ 054
济南·山东邮务管理局旧址 \ 054
济南·山东丰大银行旧址 \ 053
济南·济南电报局旧址 \ 052
威海·泰茂洋行旧址 \ 051
烟台·怡瑞兴商行旧址 \ 050
烟台·岩城商行旧址 \ 049

## 民居·府邸

济南·车站街3号原津浦铁道公司某高级职员府邸 \ 106
济南·北辛店村116号门楼 \ 105
德州·陵县徽王镇桥梁 \ 104
日照·日照港灯塔 \ 103
泰安·日观峰南泰山气象站 \ 102
泰安·南、北蓑衣亭 \ 101
泰安·风雷亭 \ 100
泰安·大众桥 \ 100
烟台·山东海关发讯台旧址 \ 099
东营·草南石桥 \ 098
青岛·栈桥 \ 097
济南·宋庄村东风桥 \ 096
济南·南桥玉符河大桥 \ 092
济南·中和桥 \ 096
济南·林泉桥 \ 091
济南·九峰桥 \ 091
济南·兴龙桥 \ 095
济南·万顺桥 \ 094
济南·原小广寒电影院 \ 095
济南·泺口黄河铁路大桥 \ 090

## 桥梁及其他

烟台·芝罘俱乐部旧址 \ 089
烟台·克利顿饭店旧址 \ 088
烟台·东太平街咖啡厅旧址 \ 087
青岛·水师饭店旧址 \ 086
青岛·青岛国际俱乐部旧址 \ 085
青岛·车站饭店旧址 \ 084

## 饭店·俱乐部

潍坊·坊子火车站德式站房 \ 083
枣庄·台儿庄火车站 \ 082
青岛·青岛火车站 \ 081
济南·万德火车站老站房 \ 080
济南·黄台车站德式建筑 \ 079

- 济南·车站街5号原津浦铁道公司某高级职员府邸 \ 107
- 济南·陈家阁 \ 108
- 济南·范家公馆 \ 109
- 济南·凤凰公馆 \ 110
- 济南·老舍旧居 \ 111
- 济南·南邓家庄恒盛门 \ 112
- 济南·彭家庄246号南门楼 \ 113
- 青岛·山头村127号绣楼 \ 113
- 济南·颜家村二区123号民居 \ 114
- 济南·原胶济铁路德国高级职员公寓 \ 115
- 济南·原金家大院 \ 116
- 济南·原金氏公馆 \ 117
- 青岛·八大关景区建筑群 \ 118
- 青岛·阿里文旧宅 \ 120
- 青岛·丛良弼公馆旧址 \ 121
- 青岛·函谷关路12号别墅 \ 122
- 青岛·花石楼 \ 123
- 青岛·龙山路18号建筑 \ 124
- 青岛·梁实秋故居 \ 124
- 青岛·陆适撰别墅 \ 125
- 青岛·湛山三路8号建筑 \ 125
- 青岛·湛山三路18号建筑 \ 126
- 青岛·湛山三路82号建筑 \ 127
- 青岛·张勋公馆旧址 \ 127
- 青岛·总督府公寓旧址 \ 128
- 东营·李青山故居 \ 128
- 烟台·衣克雷旧宅 \ 129
- 烟台·东海关职员宿舍旧址 \ 130
- 烟台·东海关总检察长官邸旧址 \ 130
- 烟台·金贡山旧宅 \ 131
- 烟台·刘子琇旧居 \ 131
- 烟台·烟台山私人别墅1号楼 \ 132
- 潍坊·金巷子6号民居 \ 133
- 泰安·田东史田纪云旧居 \ 134
- 威海·法国商人住宅旧址 \ 135
- 威海·意商别墅 \ 135
- 威海·美国牙医别墅旧址 \ 136
- 威海·英海军舰队司令避暑别墅旧址 \ 137
- 威海·英海军司令避暑房旧址 \ 138
- 威海·英商别墅旧址 \ 139
- 威海·英商私人住宅旧址 \ 140

## 教堂·庙观

- 滨州·魏氏庄园 \ 141
- 济南·洪家楼天主堂 \ 142
- 济南·白云峪村天主堂 \ 143
- 济南·北曹范村天主堂 \ 144
- 济南·陈家楼天帝庙 \ 145
- 济南·大王庙 \ 146
- 济南·东顿邱真武庙 \ 146
- 济南·后套天主堂 \ 147
- 济南·基督教自立会礼拜堂建筑群 \ 148
- 济南·济南天主教方济各会神甫修士宿舍 \ 149
- 济南·济南天主教将军庙小修院 \ 150
- 济南·三里庄基督教堂 \ 151
- 济南·万字会旧址 \ 152
- 济南·西湿口山村天主堂 \ 153
- 青岛·英国浸礼会礼拜堂 \ 154
- 青岛·江苏路基督教堂 \ 155
- 青岛·圣心修道院旧址 \ 156
- 青岛·湛山三路412号建筑 \ 157
- 青岛·湛山三路417号建筑 \ 158
- 青岛·浙江路天主堂 \ 159
- 淄博·周村基督教堂 \ 160
- 烟台·烟台联合教会旧址 \ 161
- 烟台·美国海军基督教青年会旧址 \ 162
- 泰安·卍字会旧址 \ 163
- 泰安·碧霞灵应宫 \ 164
- 泰安·无极庙 \ 165
- 泰安·小中泉天主堂 \ 166
- 泰安·新胜隆庄耶稣帝王堂 \ 167
- 泰安·玉都观 \ 168
- 威海·宽仁院旧址 \ 169
- 滨州·高庙李天主堂 \ 170
- 滨州·王家店子天主堂 \ 171
- 滨州·小刘天主堂 \ 172

# 纪念性建筑 / Commemorative Architectures

## 济南·解放阁

位于济南市历下区大明湖街道办事处黑虎泉西路（街）东端、原济南古城墙东南角、护城河东南角拐弯处。建于1948年

Located at the east end of the West Road of Heihu Spring of the Daminghu Sub-district Office of Lixia District in Jinan, or at the southeast of the ancient city wall, or at the southeast corner of the city moat. Built in 1948.

## 济南·五三惨案蔡公时殉难地

位于济南市槐荫区五里沟街道办事处经四路（街）370号。建于1920年。

Located at No.370 Jingsi Road of the Wulïgou Sub-district Office of Huaiyin District in Jinan. Built in 1920.

# 济南·五三惨案纪念碑

位于济南市历下区趵突泉街道办事处黑虎泉西路（街）西首。建于1929年

Located at the west end of the West Road of Heihu Spring of the Baotuquan Sub-district Office of Lixia District in Jinan. Built in 1929.

## 济南·辛亥革命烈士陵园

位于济南市历下区千佛山街道办事处经十一路（街）千佛山东麓。建于 1934 年

Located at the east side of the Qianto Mount Jingshiyi Road in the Qianfo Sub-district Office of Lixia District in Jinan. Built in 1934.

# 纪念性建筑

## 济南·辛稼轩纪念祠

位于济南市历下区大明湖南岸。建于1904年。

Located on the south shore of the Daming Lake in Lixia District in Jinan. Built in 1904.

## 济南·奎虚书藏

位于济南市历下区大明湖街道办事处大明湖路（街）北。建于1936年

Located on the north of Daminghu Road of the Daminghu Sub-district Office of Lixia District in Jinan. Built in 1936.

## 济南·浙闽会馆

位于济南市历下区黑虎泉西路23号。建于清末

Located at No.23 of the West Road of Heihu Spring in Lixia District in Jinan. Built in the late Qing Dynasty.

## 济南·虎豹川修路碑亭

位于济南市平阴县栾湾乡兴隆镇村。建于1929年

Located in Xinglong Village in Luanwan Town of Pingyin County in Jinan. Built in 1929.

## 济南·张氏祠堂

位于济南市历城区华山镇坝子村南部小清河北岸。建于 1921 年

Located on the north shore of Xiaoqing River, in the south of Bazi Village of Huanshan Town in Licheng District in Jinan. Built in 1921.

# 济南·广智院

位于济南市历下区广智院街。建于1905年

Located in Guangzhiyuan Street in Lixia District in Jinan. Built 1905.

### 泰安·冯玉祥读书楼

位于泰安市泰山区泰前街道环山路进贤村北 300 米。建于 1924 年
Located to the north of Jinxian Village (around 300 meters) Huanshan Road in the Taiqian Sub-district Office of Taishan District in Taian. Built in 1924.

政府·机构

## 济南·中共山东省委秘书处旧址

位于济南市天桥区制锦市街道办事处东流水街105号。建于1922年

Located at No.105 of Dongliushui Street of the Zhijinshi Sub-district Office of Tianqiao District in Jinan. Built in 1922.

## 济南·中共长清县委临时党支部成立旧址

位于济南市长清区归德镇阁楼村中。建于1938年

Located in Yanlou Village in Guide Town of Changqing County in Jinan. Built in 1938.

## 济南·中共平阴县委旧址

位于济南市平阴县孔村镇王楼村东南。建于 1939 年

Located to the southeast of Wanglou Village of Kongcun Town in Pingyin County in Jinan. Built in 1939.

## 济南·中共济南乡师支部诞生地

位于济南市天桥区北园街道办事处北园路大街364号（今明湖中学院内）。建于1929年

Located at No.364 Beiyuan Street of the Beiyuan Sub-district Office of Tianqiao District in Jinan. Built in 1929.

## 青岛·胶州帝国法院旧址

位于青岛市市南区中山路街道平度路社区德县路2号。建于1912年

Located at No.2 Dexian Road of the Pingdu Residents' Committe of the Zhongshan Sub-district Office of Shinan District in Qingdao. Built 1912.

政府·机构

## 青岛·德国总督府旧址

位于青岛市市南区中山路街道观海山社区沂水路11号。建于1903~1906年

Located at No.11 Yishui Road of the Guanhaishan Residents' Committe of the Zhongshan Street Sub-district Office of Shinan District in Qingdao. Built between 1903 and 1906.

## 青岛·德国警察署旧址

位于青岛市市南区中山路街道济南路社区湖北路 29 号。建于 1904 年

Located at No. 29 Hubei Road of the Jinan Residents' Committe of the Zhongshan Sub-district Office of Shinan District in Qingdao. Built 1904.

## 青岛·欧人监狱旧址

位于青岛市市南区江苏路街道大学路社区常州路 25 号。建于 1900 年

Located at No.25 Changzhou Road of the Daxue Residents' Committe of the Jiangsu Street Sub-district Office of Shinan District in Qingdao. Built 1900.

## 潍坊·潍坊特别市委旧址

位于潍坊市潍城区城关街道向阳路18号院内。建于1936年

Located in the yard of No. 18 Xiangyang Road of the Chengguan Sub-district Office of Weicheng District in Weifang. Built in 1936.

## 泰安·中共东平县工委诞生地

位于泰安市东平县州城镇桂井子街路东。建于民国时期

Located to the east of Guijingzi Road of Zhoucheng Town in Dongping County in Taian. Built during the period of Public of China.

政府·机构

# 临沂·山东省政府旧址

位于临沂市莒南县大店镇大店九村。建于1945年

Located in No.9 Dadian Village of Dadian Town in Junan County in Linyi. Built 1945.

# 军事设施
## Military Installations

### 济南·济南战役山东兵团指挥所旧址

位于济南市历城区仲宫镇尹家店村南部。建于 1948 年

Located to the southof Yijiadian Village of Zhonggong Town in Licheng District in Jinan. Built in 1948.

## 青岛·德国第二海军营部大楼旧址

位于青岛市市南区中山路街道观海山社区沂水路9号。建于1899年

Located at No. 9 Yishui Road of the Guanhaishan Residents' Committe of Zhongshan Sub-district Office of Shinan District in Qingdao. Built in 1899.

## 青岛·德式官邸旧址

位于青岛市市南区江苏路街道齐东路社区龙山路26号。建于1905年

Located at No.26 Longshan Road of the Qidong Residents' Committe of Jiangsu Sub-district Office of Shinan District in Qingdao. Built 1905.

## 烟台·西炮台

位于烟台市芝罘区通伸街道西炮台街 1 号（通伸岗北端山顶）。建于 1876 年

Located on the top of the North Tongshen Hill, No.1 Xipaotai Street of the Tongshen Sub-district Office of Zhifu District in Yantai. Built in 1876.

## 威海·北洋海军提督署

位于威海市环翠区鲸园街道办事处刘公岛西村丁公路。建于1887年

Located in the West Liugongdao Village of the Jingyuan Sub-district Office of Huancui District in Weihai. Built in 1887.

## 威海·东泓炮台

位于威海市环翠区鲸园街道办事处刘公岛东村。建于1892～1894年

Located in the East Liugongdao Village of the Jingyuan Sub-district Office of Huancui District in Weihai. Built between 1892 and 1894.

## 威海·华勇营大楼旧址

位于威海市环翠区鲸园街道办事处北仓居委会北山路1号院内。建于1899～1902年

Located at No.1 Beishan Road of the Beicang Residents' Committee of the Jingyuan Sub-district Office of Huancui District in Weihai. Built between 1899 and 1902.

# 威海·军官食宿处旧址

位于威海市环翠区鲸园街道办事处刘公岛西村。建于1930年前

Located r the West Liugongdao Village of the Jingyuan Sub-district Office of Huancui District in Weihai. Built in 1930.

# 临沂·八路军一一五师司令部旧址

位于临沂市莒南县大店镇大店九村。建于1942年

Located in No.9 Dadian Village of Dadian Town in Junan County in Linyi. Built 1942.

## 临沂·新四军军部旧址

位于临沂市河东区九曲街道办事处前河湾村西部。建于1946~1947年。

Located in the west of Qianhewan Village of the Jiuqu Sub-district Office in Hedong District in Linyi. Built between 1946 and 1947.

军事设施

## 临沂·八路军一一五师司令部旧址

位于临沂市莒南县大店镇大店村南北大街东侧。建于民国时期。

Located to the east of Nanbei Street in Dadian Village of Dadian Town in Junan County in Linyi. Built during the period of Public of China.

## 莱芜·莱芜战役指挥所遗址

位于莱芜市莱城区铁车乡石湾子村。建于1917年。

Located in Shiwanzi Village of Tieche Town in Laicheng District in Laiwu. Built 1917.

# 领事馆
## Consulates

## 济南·美国领事馆旧址

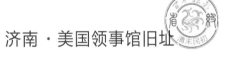

位于济南市市中区大观园街道办事处小纬二路 106 号。建于 20 世纪 20 年代

Located at No.106 Xiaoweier Road of the Dagguanyuan Sub-district Office of Shizhong District in Jinan. Built in the 1920's.

# 济南·德国领事馆旧址

位于济南市市中区大观园街道办事处经二路市政府院内。建于1901年

Located in the yard of the Municipal Government on the Jinger Road of the Daguanyuan Sub-district Office of Shizhong District in Jinan. Built in 1901.

## 济南·日本总领事馆旧址

位于济南市槐荫区五里沟街道办事处经三路240号。建于民国时期。

Located at Jingsan Road of the Wuligou Sub-district Office of Huaiyin District in Jinan. Built in the early period of Public of China.

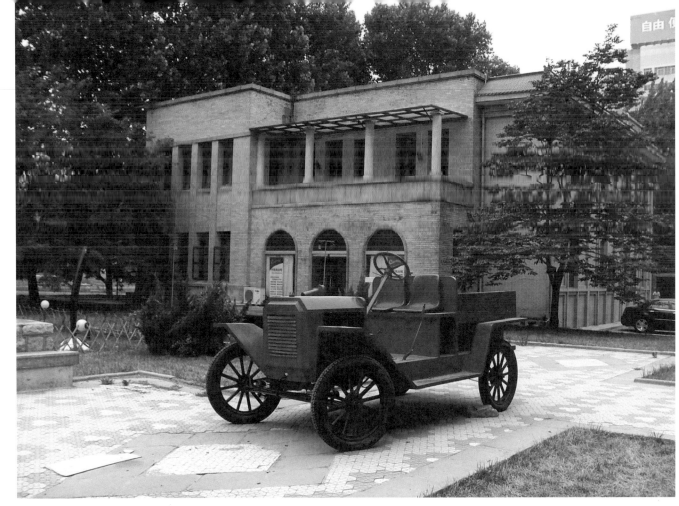

## 济南·英国领事馆旧址

位于济南市市中区大观园街道办事处经七纬四路115号。建于20世纪20年代

Located at No.115 Jingqiweisi Road of the Dagguanyuan Sub-district Office of Shizhong District in Jinan. Built in the 1920's.

## 青岛·美国领事馆旧址

位于青岛市市南区中山路街道观海山社区沂水路1号。建于1912年

Located at No.1 Yishui Road of the Guanhaishan Residents' Committe of Zhongshan Sub-district Office of Shinan District in Qingdao. Built in 1912.

## 青岛·湛山三路288号建筑

位于青岛市市北区即墨路街道恩县路社区馆陶路28号。建于1913年

Located at No.28 Guantao Road of Enxian Residents' Committe of the Jimo Sub-district Office of Shibei District in Qingdao. Built in 1913.

## 烟台·丹麦领事馆旧址

位于烟台市芝罘区向阳街街道办事处烟台山西路街15号。建于1890年

Located at No.15 West Yantaishan Road of the Xiangyang Sub-district Office of Zhifu District in Yantai. Built in 1890.

## 烟台·俄国领事馆旧址

位于烟台市芝罘区东山街道办事处大马路街108号。建于1904年。

Located at No.108 Damalu Street of the Dongshan Sub-district Office of Zhifu District in Yantai. Built in 1904.

# 烟台·美国领事馆领事官邸旧址

位于烟台市芝罘区向阳街道办事处历新路街7号。建于1900年

Located at No.7 Lixin Road of the Xiangyang Sub-district Office of Zhifu District in Yantai. Built in 1900.

## 烟台·挪威领事馆旧址

位于烟台市芝罘区向阳街道办事处海岸街9号。建于1904年

Located at No.9 Haian Road of the Xiangyang Sub-district Office of Zhifu District in Yantai. Built in 1904.

## 烟台·日本领事馆旧址

位于烟台市芝罘区向阳街道办事处烟台山西路街4号。建于1938年

Located at No.4 West Yantaishan Road of the Xiangyang Sub-district Office of Zhifu District in Yantai. Built in 1938.

## 烟台·英国领事馆附属建筑旧址

位于烟台市芝罘区向阳街道办事处烟台山东路街。建于19世纪中叶

Located in East Yantaishan Road of the Xiangyang Sub-district Office of Zhifu District in Yantai. Built in the middle 19th century.

# 商行·商号

Trading companies and Firms

复成信东记

复成信西记

## 济南·复成信东记、西记

位于济南市槐荫区五里沟街道办事处经二路（街）363号。建于1909年

Located at No.363 Jinger Road of the Wuligou Sub-district Office of Huaiyin District in Jinan. Built in 1909.

## 济南·宏济堂西记

位于济南市槐荫区五里沟街道办事处原经二路（街）295号。建于民国时期

Located at No.295 the former Jinger Road of the Wuligou Sub-district Office of Huaiyin District in Jinan. Built in the early period of Public of China.

## 济南·瑞蚨祥鸿记旧址

位于济南市市中区大观园街道办事处经二路（街）215号。建于1923年

Located at No.215 Jinger Road of the Dagguanyuan Sub-district Office of Shizhong District in Jinan. Built in 1923.

## 济南·原皇宫照相馆

位于济南市市中区大观园街道办事处经一路（纬）108号。建于民国初期

Located at No.108 Jinger Road of the Dagguanyuan Sub-district Office of Shizhong District in Jinan. Built in the early period of Public of China.

## 青岛·礼和商业大楼旧址

位于青岛市市南区中山路街道单县路社区太平路41号。建于1902年

Located at No. 418 Taiping Road of the Shanxian Residents' Committe of Zhongsahn Sub-district Office of Shinan District in Qingdao. Built in 1902.

## 青岛·湛山三路 287 号建筑

位于青岛市市北区即墨路街道恩县路社区馆陶路 37 号。建于 1927 年

Located at No.37 Guantao Road of the Enxi-an Residents' of Committe Jimo Sub-district Office of Shibei District in Qingdao. Built in 1927.

## 淄博·仁德茶庄旧址

位于淄博市周村区大街街道办事处长安社区大街南段路西101号。建于民国时期。

Located at No.101 to the west of the south Changan Street of the Dajie Sub-district Office of Zhoucun District in Zibo. Built in the period of Public of China.

## 淄博·瑞蚨祥旧址

位于淄博市周村区永安街道办事处朝阳社区大车馆街1号。建于民国时期

Located at No.1 Dacheguan Street of the Chaoyang Residents' Committe of the Yongan Sub-district Office of Zhoucun District in Zibo. Built in the period of Public of China.

## 烟台·法商永兴洋行旧址

位于烟台市芝罘区向阳街道办事处海岸街20号。建于民国时期

Located at No.20 Haian Road of the Xiangyang Sub-district Office of Zhifu District in Yantai. Built in the period of Public of China.

## 烟台·庆昌五金行旧址

位于烟台市芝罘区向阳街道办事处顺太街8号。建于1920年

Located at No.8 Shuntai Road of the Xiangyang Sub-district Office of Zhifu District in Yantai. Built in 1920.

## 烟台·顺昌商行旧址

位于烟台市芝罘区向阳街道办事处朝阳街44号。建于民国时期

Located at No.44 Chaoyang Road of the Xiangyang Sub-district Office of Zhifu District in Yantai. Built in the period of Public of China.

## 烟台·万国理发馆旧址

位于烟台市芝罘区向阳街道办事处建德街23-24号。建于民国时期

Located at No.23-24 Daode Road of the Xiangyang Sub-district Office of Zhifu District in Yantai. Built in the period of Public of China.

## 烟台·五洲大药房旧址

位于烟台市芝罘区向阳街道办事处朝阳街 60-61 号。建于民国时期

Located at No.60-61 Chaoyang of the Xiangyang Sub-district Office of Zhifu District in Yantai. Built in the period of Public of China.

## 烟台·岩城商行旧址

位于烟台市芝罘区向阳街道办事处顺太街15号。建于民国时期

Located at No.15 Shuntai Road of the Xiangyang Sub-district Office of the Zhifu District in Yantai. Built in the period of Public of China.

# 烟台·怡瑞兴商行旧址

位于烟台市芝罘区向阳街道办事处顺泰街16号。建于民国时期

Located at No.165 Shuntai Road of the Xiangyang Sub-district Office of the Zhifu District in Yantai. Built in the period of Public of China.

商行·商号

## 威海·泰茂洋行旧址

位于威海市环翠区鲸园街道办事处东山居委会军港内。建于1919年

Located in the Army Harbor of the Dongshan Residents' Committee of the Jingyuan Sub-district Office of Huancui District in Weihai. Built in 1919.

# 银行·邮局

Banks and Post Offices

## 济南·济南电报局旧址

位于济南市天桥区纬北路街道办事处经一路（街）91号。建于1904年

Located at No.91 Jingyi Road of the Weibei Sub-district Office of Tianqiao District in Jinan. Built in 1904.

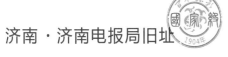

银行·邮局

## 济南·山东丰大银行旧址

位于济南市槐荫区五里沟街道办事处纬六路（街）27号。建于1919年

Located at No.27 Weiliu Road of the Wuligou Sub-district Office of Huaiyin District in Jinan. Built in 1919.

原门头字匾

## 济南·山东邮务管理局旧址

位于济南市市中区大观园街道办事处经二路（街）158号。建于1919年

Located at No.158 Jinger Road of the Dagguanyuan Sub-district Office of Shizhong District in Jinan. Built in 1919.

## 青岛·大陆银行旧址

位于青岛市市南区中山路街道中山路社区中山路70号。建于1934年

Located at No.70 Zhongsahn Road of the Zhongshan Residents' Committe of the Zhongshan Sub-district Office of Shinan District in Qingdao. Built 1934.

# 青岛·德华银行旧址

位于青岛市市南区中山路街道浙江路社区广西路 14 号。建于 1899～1901 年

Located at No.14 Guangxi Road of the Zhejiang Residents' Committe of the Zhongshan Sub-district Office of Shinan District in Qingdao. Built between 1899 and 1901.

## 青岛·帝国邮政局旧址

位于青岛市市南区中山路街道浙江路社区安徽路5号。建于1901年

Located at No.5 Anhui Road of the Zhejiang Residents' Committe of the Zhongshan Sub-district Office of Shi-nan District in Qingdao. Built in1901.

银行・邮局

## 青岛・交通银行青岛分行旧址

位于青岛市市南区中山路街道中山路社区中山路93号。建于1931年
Located at No.93 Zhongsahn Road of the Zhoangshan Residents' Committe of the Zhongshan Sub-district Office of Shinan District in Qingdao. Built 1931.

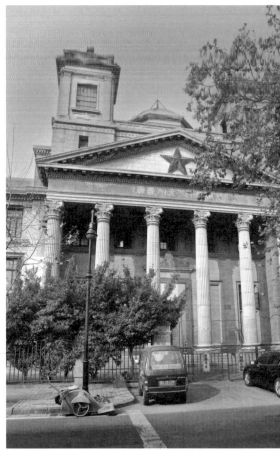

## 青岛·湛山三路 290 号建筑

位于市北区即墨路街道恩县路社区馆陶路 22 号。建于 1925 年

Located at No.22 Guantao Road of the Enxian Residents' Committe Jimo of the Sub-district Office of Shibei District in Qingdao. Built in 1925.

# 青岛·湛山三路293号建筑

位于青岛市市北区即墨路街道吴淞路社区馆陶路5号。建于1917年

Located at No.5 Guantao Road of the Wusong Residents' Committe of the Jimo Sub-district Office of Shibei District in Qingdao. Built in 1917.

## 烟台·德国邮局旧址

位于烟台市芝罘区向阳街道办事处海岸街 18 号。建于 1892 年

Located at No.18 Haian Road of the Xiangyang Sub-district Office of Zhifu District in Yantai. Built in 1892.

## 烟台·汇丰银行烟台分行旧址

位于烟台市芝罘区向阳街道办事处海关街17号。建于1920年

Located at No.17 Haiguan Road of the Xiangyang Sub-district Office of Zhifu District in Yantai. Built in 1920.

## 烟台·交通银行烟台支行旧址

位于烟台市芝罘区向阳街道办事处海关街 29 号。建于 1910 年

Located at No.29 Haiguan Road of the Xiangyang Sub-district Office of Zhifu District in Yantai. Built in 1910.

银行·邮局

## 烟台·烟台邮局旧址

位于烟台市芝罘区向阳街道办事处海岸街29号。建于1925年

Located at No.29 Haian Road of the Xiangyang Sub-district Office of Zhifu District in Yantai. Built in 1925.

# 学校 · 医院
Educational Institutions and Hospitals

## 济南 · 原济南市第一中学教学楼

位于济南市历下区大明湖街道办事处运署街 43 号。建于 1903 年

Located at No.43 Yunshu Street in the Daminghu Sub-district Office of Lixia District in Jinan. Built in 1903.

柏根楼

共和楼

考文楼

求真楼

## 济南·原齐鲁大学近现代建筑群

位于济南市历下区趵突泉街道办事处文化西路（街）山东大学（原山东医科大学）院内。建于 1917 年

Located in the yard of Shandong University (the former Shandong Medical University) on Wenhuaxi Road of the Baotuquan Sub-district Office of Lixia District in Jinan. Built in 1917.

## 济南·懿范女子中学

位于济南市历城区洪家楼5号。建于1936年

Located at No.5 Hongjialou Village of Licheng District in Jinan. Built in 1936.

学校·医院

## 济南·正谊中学

位于济南市历下区明湖路路北。建于20世纪20年代

Located to the north of Minghu Road of Lixia District in Jinan. Built in the 1920's.

## 济南·山东省红卍字会施诊所

位于济南市市中区魏家庄民康里6号。建于1928年。

Located at No.6 Minkangli Residents' Committe of the Weijiazhuang Sub-district Office of Shizhong District in Jinan. Built in 1928.

## 济南·同仁会济南医院旧址

位于济南市槐荫区五里沟街道办事处经五路（街）324号。建于1917年

Located at No.324 Jingwu Road of the Wuligou Sub-district Office of Huaiyin District in Jinan. Built in 1917.

## 淄博·光被中学旧址

位于淄博市周村区大街街道办事处和平社区育才路2号。建于1909年

Located at No.2 Yucai Road in the Heping Residents' Committe of the Dajie Sub-district Office of Zhoucun District in Zibo. Built in 1909.

## 烟台·崇正中学旧址

位于烟台市芝罘区向阳街道大马路街109号。建于1931年

Located at No.109 Damalu Road of the Xiangyang Sub-district Office of Zhifu District in Yantai. Built in 1931.

## 烟台·市立烟台医院旧址

位于烟台市芝罘区向阳街道办事处海岸街20号。建于1934年

Located at No.20 Haian Road of the Xiangyang Sub-district Office of Zhifu District in Yantai. Built in 1934.

## 烟台·益文商业专科学校旧址

位于烟台市芝罘区毓璜顶街道办事处焕新路街1号。建于1920年

Located at No.1 Hunxin Road of the Yuhuangding Sub-district Office of Zhifu District in Yantai. Built in 1920.

## 潍坊·乐道院医院十字楼

位于潍坊市奎文区广文街道李家村西 200 米、市人民医院北办公及生活区内。建于 1924 年

Located around 200 meters to the west of Lijia Village of the Guangwen Sub-district Office of Kuiwen District in Weifang, in the living quarters and working sections to the north of The municipal people's hospital. Built in 1924.

## 威海·威海卫学校旧址

位于威海市环翠区孙家疃镇黄泥沟村环海路 5 号。建于 1903～1925 年

Located at No.5 Huanhai Road in Huangnigou Village of Sunjiatong Town of Huancui District in Weihai. Built between 1903 and 1925.

学校·医院

## 滨州·英国教会医院旧址

位于滨州市惠民县孙武镇南关街环城南路108号。建于1919年。

Located at No.108 the south Huancheng Road Nanguan Street of Sunwu Town of Huimin County in Binzhou. Built in 1919.

## 临沂·美国教会医院旧址

　　位于临沂市解放路东段 27 号。建于清末

Located at No.27 in the east end of Jiefang Road in Linyi.
Built in the late Qing Dynasty.

## 济南·济南火车站

位于济南市天桥区车站街19号。建于1904年

Located at No.19 Chezhan Road of Tianqiao District in Jinan. Built in 1904.

车站 Stations

## 济南·北关车站旧址

位于济南市天桥区北坦街道办事处北关北路（街）角楼庄 30 号。建于 1930 年

Located at No.30 Juelou Village of the north Beiguan Street of the Beitan Sub-district Office of Tianqiao District in Jinan. Built in 1930.

车 站

### 济南·黄台车站德式建筑

位于济南市历下区东关街道办事处山大路北端。建于1905年

Located at the north end of Shanda Road of the Dongguan Sub-district Office of Lixia District in Jinan. Built in 1905.

### 济南·万德火车站老站房

位于济南市长清区万德镇万北村东万德火车站内。建于1904年

Located in the Wande Station in Wanbei Village of Wande Town of Changqing District in Jinan. Built in 1904.

## 青岛·青岛火车站

位于青岛市市南区泰安路。建于 1900 年

Located on Taian Road of Shinan District in Qingdao. Built in 1900.

## 枣庄·台儿庄火车站

位于枣庄市台儿庄。建于1899年

Located in Taierzhuang in Zaozhuang. Built in 1899.

## 潍坊·坊子火车站德式站房

　　位于潍坊市坊子区坊城街道城社区一马路东段路北、铁路线以南。建于 1902 年

Located to the north of East No.1 Road and to the south of the railway Daocheng Residents' Committe of the Fangcheng Sub-district Office of Fangzi District in Weifang. Built in 1902.

# 饭店·俱乐部

Hotels and Clubs

## 青岛·车站饭店旧址

位于青岛市市南区中山路街道单县路社区兰山路28号。建于1913年

Located at No.28 Lanshan Road of the Shanxian Residents' Committe of the Zhongshan Street Sub-district Office of Shinan District in Qingdao. Built 1913.

## 青岛·青岛国际俱乐部旧址

位于青岛市市南区中山路街道中山路社区中山路1号。建于1910年

Located at No.1 Zhongshan Road of the Zhongshan Residents' Committte of the Zhongshan Sub-district Office of Shinan District in Qingdao. Built 1910.

## 青岛·水师饭店旧址

位于青岛市市南区中山路街道中山路社区湖北路17号乙。建于1901～1902年

Located at No.17 Hubei Road of the Zhongshan Residents' Committe of the Zhongshan Sub-district Office of Shinan District in Qingdao. Built between 1901 and 1902.

## 烟台·东太平街咖啡厅旧址

位于烟台市芝罘区向阳街道办事处东太平街。建于民国时期

Located in Dongtaiping Road of the Xiangyang Sub-district Office of Zhifu District in Yantai. Built during the period of Public of China.

## 烟台·克利顿饭店旧址

位于烟台市芝罘区向阳街道办事处朝阳街 44 号。建于 1910 年

Located at No.44 Chaoyang Road of the Xiangyang Sub-district Office of Zhifu District in Yantai. Built in 1910.

饭店·俱乐部

## 烟台·芝罘俱乐部旧址

位于烟台市芝罘区向阳街道海岸街 34 号。建于 1865 年

Located at No.34 Haian Road of the Xiangyang Sub-district Office of Zhifu District in Yantai. Built in 1865.

# 桥梁及其他

Bridges and others

## 济南·泺口黄河铁路大桥

位于济南市泺口街道。建于1909年

Located in the Luokou Sub-district Office in Jinan. Built in 1909.

## 济南·九峰桥

位于济南市历城区彩石镇东泉村中。建于 1919 年

Located in Dongquan Village of Caishi Town of Licheng District in Jinan. Built in 1919.

## 济南·林泉桥

位于济南市历城区仲宫镇锦绣川办事处北坡村南、南靠南庄。建于 1941 年

Located to the south of Beipo Village and to the north of the Nanzhuang of the Jinxiuchuan District Office of Zhonggong Town of Licheng District in Jinan. Built in 1941.

## 济南·南桥玉符河大桥

位于济南市长清区平安街道办事处南桥村北。建于1933年

Located in the north of Nanqiao Village of the Pingan Sub-district Office of Changqing County in Jinan. Built in 1933.

## 济南·宋庄村东风桥

位于济南市平阴县安城乡宋庄村北。建于民国时期

Located to the north of Songzhuang Village of Ancheng Town of Pingyin County in Jinan. Built during the period of Public of China.

## 济南·太和桥

位于济南市历城区遥墙镇大陈家村东约500米处巨野河上。建于民国时期

Located on the Juye River, around 500 meters to the east of Dachenjia Village of Yaoqiang Town of Licheng District in Jinan. Built during the period of Public of China.

## 济南·万顺桥

位于济南市历城区彩石镇白腊滩村。建于1945年

Located in Bailatan Village of Caishi Town of Licheng District in Jinan. Built in 1945.

桥梁及其他

## 济南·兴龙桥

位于济南市历城区遥墙镇陈家岭村东北巨野河上。建于民国时期

Located on the Juye River, to the northeast of Chenjialing Village of Yaoqiang Town of Licheng District in Jinan. Built during the period of Public of China.

## 济南·原小广寒电影院

位于济南市市中区大观园街道办事处经三路48号。建于1914年　　Located at No. 48 Jingsan Road of the Dagguanyuan Sub-district Office of Shizhong District in Jinan. Built in 1914.

## 济南·中和桥

位于济南市历城区港沟镇伙路村中部。
建于 1922 年

Located in the middle of Huolu Village of Ganggou Town of Licheng District in Jinan. Built in 1922.

## 青岛·栈桥

位于青岛市市南区太平路 12 号。建于 1891 年

Located at No.12 Taiping Road of Shinan District in Qingdao. Built in 1891.

## 东营·草南石桥

位于东营市广饶县花官乡草南村北。建于1946年

Located to the north of Caonan Village of Huaguan Town of Guangrao County in Dongying. Built in 1946.

## 烟台·山东海关发讯台旧址

位于烟台市芝罘区向阳街道烟台山西路街 26 号。建于 1933 年

Located at No.26 the west Yantaishan Road of the Xiangyang Sub-district Office of Zhifu District in Yantai. Built in 1933.

## 泰安·大众桥

位于泰安市泰山区环山路北冯玉祥墓之西。建于民国时期

Located to the west of Fengyuxiang Grave and to the north of Huanshan Road of Taishan District in Taian. Built during the period of Public of China.

## 泰安·风雷亭

位于泰安市泰山区泰山长寿桥西。建于民国时期

Located to the west of Taishan Long-live Bridge of Taishan District in Taian. Built during the period of Public of China.

## 泰安·南、北蓑衣亭

位于泰安市泰山区泰山环山路北首大众桥南、北两侧。建于民国时期

Located to the north of Dazhong Bridge, at the north end of Huanshan Road of Taishan District in Taian. Built during the period of Public of China.

## 泰安·日观峰南泰山气象站

位于泰安市泰山区泰山之巅的日观峰上。建于民国时期

Located at Riguanfeng on the top of Taishan of Taishan District in Taian. Built during the period of Public of China.

## 日照·日照港灯塔

位于日照市东港区石臼街道三村日照港务局一公司食堂东侧。建于1933年

Located to the east of the dining room belonged to the First Company of Rizhao Harbour Bureau, Sancun of the Shijiu Sub-district Office of Donggang District in Rizhao. Built in 1933.

## 德州·陵县徽王镇桥梁

位于陵县徽王镇以东大约一公里的马颊河故道上。建于民国时期

Located on the old course of Majia River, around one kilometer to the east of Huiwan Twon in Ling County. Built during the period of Public of China.

# 民居·府邸

Folk houses and Mansions

## 济南·北辛店村 116 号门楼

位于济南市历城区华山镇北辛店村东南部 116 号。建于民国时期。

Located at No.116 in the southeast Beixindian Village of Huashan Town of Licheng District in Jinan. Built during the period of Public of China.

## 济南·车站街 3 号原津浦铁道公司某高级职员府邸

位于济南市天桥区纬北路街道办事处车站街 3 号。建于 20 世纪初

Located at No.3 Chezhan Road of the Weibei Sub-district Office of Tianqiao District in Jinan. Built in the early 20th century.

## 济南·车站街 5 号原津浦铁道公司某高级职员府邸

位于济南市天桥区纬北路街道办事处车站街 5 号。建于 20 世纪初

Located at No.5 Chezhan Road of the Weibei Street Sub-district Office of Tianqiao District in Jinan. Built in the early 20th century.

## 济南·陈家阁

位于济南市历城区遥墙镇鸭旺口村村东、镇粮所院内。建于 1945 年

Located in the yard of the Towan Grain Depot, to the east of Yawanglou Village of Yaoqiang Town of Licheng District in Jinan. Built in 1945.

## 济南·范家公馆

位于济南市长清区万德镇小万德村东。建于民国时期

Located to the east of Xiaowande Village of Wande Town of Changqing District in Jinan. Built during the period of Public of China.

## 济南·凤凰公馆

位于济南市历城区港沟镇（现属临港开发区）凤鸣路（街）1000号。建于1910年前后

Located at No.1000 Fengming Road Ganggou Town (now belonging to the Lingang Development Zone) of Licheng District in Jinan. Built around 1910.

## 济南·老舍旧居

位于济南市历下区趵突泉街道办事处南新街58号。建于20世纪30年代

Located at No.58 Nanxin Road of the Baotuquan Sub-district Office of Lixia District in Jinan. Built in the 1930's.

## 济南·南邓家庄恒盛门

位于济南市章丘市曹范镇南邓家庄村中西北部。建于民国时期。

Located in the middle northwest of Nandengjia Village of Caofan Town of Zhangqiu in Jinan. Built during the period of Public of China.

民居·府邸

## 济南·彭家庄 246 号南门楼

位于济南市历城区郭店镇彭家村南 246 号。建于民国时期

Located at No.246, to the south of Pengjia Village of Guodian Town of Licheng District in Jinan. Built during the period of Public of China.

## 济南·山头村 127 号绣楼

位于济南市历城区郭店镇山头村中部。建于民国时期

Located in the middle of Shantou Village of Guodian Town of Licheng District in Jinan. Built during the period of Public of China.

## 济南·颜家村二区 123 号民居

位于济南市历城区唐王镇颜家村中街西头二区 123 号。
建于 1932 年

Located at No.123, the Second Section of the west Zhongjie of Yanjia Village of Tangwang Town of Licheng District in Jinan. Built in 1932.

民居·府邸

## 济南·原胶济铁路德国高级职员公寓

位于济南市市中区大观园街道办事处经一纬二路（街）19号。建于1915年

Located at No.19 Jingyiweier Road of the Dagguanyuan Sub-district Office of Shizhong District in Jinan. Built in 1915.

## 济南·原金家大院

位于济南市历下区大明湖街道办事处宽厚所街 55 号。建于民国时期

Located at No.55 Kuanhousuo Road of the Daminghu Sub-district Office of Lixia District in Jinan. Built during the period of Public of China.

## 济南·原金氏公馆

位于济南市市中区泺源街道办事处麟趾巷 30 号。建于 1933 年

Located at No.30 Linzhi Road of the Luoyuan Sub-district Office of Shizhong District in Jinan. Built in 1933.

# 青岛·八大关景区建筑群

位于青岛市太平角汇泉岬拱卫左右。建于民国时期

Located at Taipingjiao in Qingdao, Huiquan Capes standing besides.
Built during the period of Public of China.

民居・府邸

## 青岛·阿里文旧宅

位于青岛市市南区八大关街道鱼山路社区鱼山路1号。建于1899年

Located at No.1 Yushan Road of the Yushan Residents' Committee of the Badaguan Sub-district Office of Shinan District in Qingdao. Built in 1899.

民居·府邸

## 青岛·丛良弼公馆旧址

位于青岛市市南区江苏路街道齐东路社区齐东路2号。建于1925年

Located at No.2 Qidong Road of the Qidong Residents' Committee of the Jiangsu Street Sub-district Office of Shinan District in Qingdao. Built in 1925.

## 青岛·函谷关路 12 号别墅

位于青岛市市南区八大关街道八大关社区函谷关路 12 号。建于 1946 年

Located at No.12 Hanguguan Road of the Badaguan Residents' Committee of the Badaguan Sub-district Office of Shinan District in Qingdao. Built in 1946.

## 青岛·花石楼

位于青岛市八大关风景疗养区黄海路18号。建于1930年

Located at No.18 Huanghai Road, in the Badaguan Landscape convalescence area in Qingdao. Built in 1930.

## 青岛·龙山路 18 号建筑

位于青岛市市南区江苏路街齐东路社区龙山路 18 号。建于 1933 年

Located at No.18 Longsahn Road of the Qidong Residents' Committee of the JIangsu Sub-district Office of Shinan District in Qingdao. Built in 1933.

## 青岛·梁实秋故居

位于青岛市市南区八大关街道福山路社区鱼山路 33 号。建于 1928 年

Located at No.33 Yushan Road of the Fushan Residents' Committee of the Badaguan Sub-district Office of Shinan District in Qingdao. Built in 1928.

民居・府邸

### 青岛·陆适撰别墅

位于青岛市市南区八大关街道八大关社区函谷关路30号。建于1936年

Located at No.30 Hanguguan Road of the Badaguan Residents' Committee of the Badaguan Sub-district Office of Shinan District in Qingdao. Built in 1936.

### 青岛·湛山三路8号建筑

位于青岛市市南区八大关街道太平角社区湛山三路6号。建于1941年

Located at No.6 Zhanshansan Road of the Badaguan Residents' Committee of the Taipingjiao Sub district Office of Shinan District in Qingdao. Built in 1941.

## 青岛·湛山三路 18 号建筑

位于青岛市市南区八大关街道八大关社区嘉峪关路 17 号。建于 1935 年

Located at No.17 Jiayuguan Road of the Badaguan Residents' Committee of the Badaguan Sub-district Office of Shinan District in Qingdao. Built in 1935.

## 青岛·湛山三路 82 号建筑

位于青岛市市南区八大关街道八大关社区居庸关路 10 号。建于 1941 年

Located at No.10 Juyongguan Road of the Badaguan Residents' Committee of the Badaguan Sub-district Office of Shinan District in Qingdao. Built in 1941.

## 青岛·张勋公馆旧址

位于青岛市市南区浙江路 7 号。建于 1900 年

Located at No.7 Zhejiang Road of Shinan District in Qingdao. Built in 1900.

## 青岛·总督府公寓旧址

位于青岛市市南区中山路街道观海山社区沂水路3号。建于1901年

Located at No.3 Yishui Road of the Guanhaishan Residents' Committee of the Zhongshan Sub-district Office of Shinan District in Qingdao. Built in 1901.

## 东营·李青山故居

位于东营市广饶县稻庄镇稻二村东。建于民国时期

Located to the east of Daoer Village of Daozhuang Town of Guangrao County in Dongying. Built during the period of Public of China.

## 烟台·东海关职员宿舍旧址

位于烟台市芝罘区向阳街道办事处烟台山西路街28-32号。建于1904年

Located at No.28-32 West Yantaishan Road of the Xiangyang Sub-district Office of Zhifu District in Yantai. Built in 1904.

## 烟台·东海关总检察长官邸旧址

位于烟台市芝罘区向阳街道烟台山西路街 25 号。建于 19 世纪中叶

Located at No.25 West Yantaishan Road of the Xiangyang Sub-district Office of Zhifu District in Yantai. Built in the middle 19th century.

## 烟台·金贡山旧宅

位于烟台市芝罘区东山街道大马路街 60 号。建于 1937 年

Located at No.60 Damalu Road of the Dongshan Sub-district Office of Zhifu District in Yantai. Built in 1937.

## 烟台·刘子琇旧居

位于烟台市芝罘区向阳街道办事处向阳街 81 号。建于 1920 年

Located at No.81 Xiangyang Road of the Xiangyang Sub-district Office of Zhifu District in Yantai. Built in 1920.

## 烟台·烟台山私人别墅 1 号楼

位于烟台市芝罘区向阳街道办事处办新路街 7 号。建于 19 世纪末

Located at No.7 Lixin Road of the Xiangyang Sub-district Office of Zhifu District in Yantai. Built in the late 19th century.

## 烟台·衣克雷旧宅

位于烟台市芝罘区毓璜顶街道办事处焕新路街 3 号。建于 20 世纪 40 年代

Located at No.3 Hunxin Road of the Yuhuangding Sub-district Office of Zhifu District in Yantai. Built in the 1940's.

民居·府邸

## 潍坊·金巷子 6 号民居

位于潍坊市潍城区城关街道金巷子 6 号。建于 20 世纪 20 年代

Located at No.6 Jinxiangzi of the Chengguan Sub-district Office of Weicheng District in Weifang. Built in the 1920's.

## 泰安·田东史田纪云旧居

位于泰安市肥城市汶阳镇田东史村。建于民国时期

Located in Tiandongshi Village of Wenyang Town of Feicheng City inTaian. Built during the period of Public of China.

民居·府邸

## 威海·法国商人住宅旧址

位于威海市环海路10号。建于1932年

Located at No.10 Huanhai Road in Weihai. Built in 1932.

## 威海·美国牙医别墅旧址

位于威海市环翠区孙家疃镇黄泥沟村环海路九号。建于1931年

Located at No.9 Huanhai Road in Huangnigou Village of Sunjiatong Town of Huancui District in Weihai. Built in 1931.

## 威海·意商别墅旧址

位于威海市环翠区孙家疃镇黄泥沟村环海路9号。建于1920年。

Located at No.9 Huanhai Road in Huangnigou Village of Sunjia-Tong Town of Huancui District in Weihai. Built in 1920.

民居·府邸

## 威海·英海军舰队司令避暑别墅旧址

位于威海市环翠区鲸园街道办事处刘公岛东村。建于 1898 年

Located in the Eest Liugongdao Village of the Jingyuan Sub-district Office of Huancui District in Weihai. Built in 1898.

## 威海·英海军司令避暑房旧址

位于威海市环翠区孙家疃镇黄泥沟村环海路7-1号。建于1904年

Located at No.7-1 the Eest Liugongdao Village of the Jingyuan Sub-district Office of Huancui District in Weihai. Built in 1904.

## 威海·英商别墅旧址

位于威海市环翠区孙家疃镇黄泥沟村环海路11号中国人民解放军威海职工疗养院院内。建于1934～1936年

Located at No.11 the Eest Liugongdao Village of the Jingyuan Sub-district Office of Huancui District in Weihai (in the yard of PLA Weihai Staff Sanatorium). Built in 1934 and 1936.

## 威海·英商私人住宅旧址

位于威海市环翠区鲸园街道办事处东山居委会东山路18号。建于1913年

Located at No.18 Dongshan Road of the Dongshan Residents' Committee of the Jingyuan Sub-district Office of Huancui District in Weihai. Built 1913.

民居·府邸

## 滨州·魏氏庄园

位于滨州市惠民县县城东南部的魏集镇。建于1886年

Located in Weiji Town, to the southeast of the city of Huimin County in Binzhou. Built in 1886.

# 教堂 · 庙观
## Churches and Temples

### 济南 · 洪家楼天主堂

位于济南市历城区洪家楼街道办事处（镇）洪家楼村北路 1 号院内。建于 1906 年

Located in the yard of No.1 Beilu of Hongjialou Village of the Hongjialou Sub-district Office of Licheng District in Jinan. Built in 1906.

## 济南·白云峪村天主堂

位于济南市平阴县孔村镇白云峪村中。建于 1893 年

Located in Baiyunyu Village of Kongcun Town of Pingyin County in Jinan. Built in 1893.

## 济南·北曹范村关帝庙

位于济南市章丘市曹范镇北曹范村卫生所后院。建于民国时期

Located in the back yard of the Beicaofan Clinic of Caofan Town of Zhangqiu in Jinan. Built during the period of Public of China.

## 济南·陈家楼天主堂

位于济南市天桥区纬北路街道办事处前陈家楼63号［原北坦南路（街）3号］。建于1908年

Located at No.63 the front Chenjialou Village of the Weibei Sub-district Office of Tianqiao District in Jinan(originally: No. 3 the south Beitan Road). Built in 1908.

## 济南·大王庙

位于济南市历城区柳埠镇柳东村东北部。建于民国时期

Located in the northeast of Liudong Village of Liubu Town of Licheng District in Jinan. Built during the period of Public of China.

## 济南·东顿邱真武庙

位于济南市历城区孙村镇东顿邱村偏北。建于民国时期

Located to the bit north of Dongdunqiu Village of Suncun Town of Licheng District in Jinan. Built during the period of Public of China.

## 济南·后套天主堂

位于济南市平阴县孔村镇后套村北。建于1913年

Located to the north of Houtao Village of Kongcun Town of Pingyin County in Ji'nan. Built in 1913.

## 济南·基督教自立会礼拜堂建筑群

位于济南市槐荫区五里沟街道办事处经四路 425 号。建于 1926 年

Located at No.425 Jingsi Road of the Wuligou Sub-district Office of Huaiyin District in Jinan. Built in 1926.

## 济南·济南天主教方济各会神甫修士宿舍

位于济南市历城区洪家楼花园路东首。建于 1932 年

Located at the east end of Huayuan Road of the Hongjialou Sub-district Office of Licheng District in Jinan. Built in 1932.

## 济南·济南天主教将军庙小修院

位于济南市历下区将军庙街路北71号。建于1863年

Located at No.71 to the north of Jiangjunmiao Road of Lixia District in Jinan. Built in 1863.

## 济南·三里庄基督教堂

位于济南市市中区大观园街道办事处三里庄路（街）13号。建于民国时期

Located at No.13 Sanlizhaung Road of the Daguanyuan Sub-district Office of Shizhong District in Jinan. Built during the period of Public of China.

## 济南·万字会旧址

位于济南市市中区泺源街道办事处上新街51号。建于1934年

Located at No.51 Shangxin Road of the Luoyuan Sub-district Office of Shizhong District in Jinan. Built in 1934.

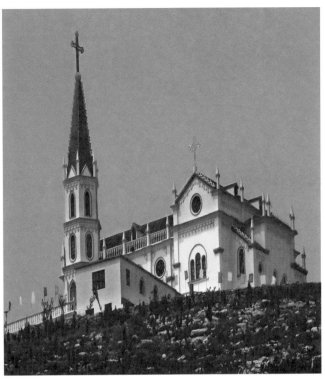

## 济南·西湿口山村天主堂

位于济南市平阴县孝直镇西湿口山村西北。建于 1908 年

Located to the northwest of Xishishankou Village of Xiaozhi Town of Pingyin County in Jinan. Built in 1908.

## 济南·英国浸礼会礼拜堂

位于济南市历下区趵突泉街道办事处广智（院）街 76 号。建于 1905 年

Located at No.76 Guangzhiyuan Road of the Baotuquan Sub-district Office of Lixia District in Jinan. Built in 1905.

## 青岛·江苏路基督教堂

位于青岛市市南区中山路街道观海山社区江苏路15号。建于1908～1910年

Located at No.15 Jiangsu Road of the Guanhaishan Residents' Committee of the Zhongshan Sub-district Office of Shinan District in Qingdao. Built between 1908 and 1910.

## 青岛·圣心修道院旧址

位于青岛市市南区中山路街道中山路社区浙江路 28 号。建于 1901 年

Located at No.28 Zhejiang Road of the Zhongshan Residents' Committee of the Zhongshan Sub-district Office of Shinan District in Qingdao. Built in 1901.

## 青岛·湛山三路412号建筑

位于即墨市环秀街道办事处河南信义街105、107号。建于1918年

Located at No.105 and 107 Henanxinyi Road of the Huanxiu Sub-district Office in Jimo. Built in 1918.

## 青岛·湛山三路 417 号建筑

位于青岛市即墨市潮海街道办事处平等街 52 号。建于 1924 年

Located at No.52 Pingdeng Road of the Chaohai Sub-district Office of Jimo City in Qingdao. Built in 1924.

教堂·庙观

# 青岛·浙江路天主堂

位于青岛市市南区中山路街道中山路社区浙江路 15 号。建于 1932 年

Located at No.15 Zhejiang Road of the Zhongshan Residents' Committe of the Zhongshan Sub-district Office of Shinan District in Qingdao. Built 1932.

## 淄博·周村基督教堂

位于淄博市周村区丝市街663号。建于1911年

Located at No.663 Sishi Road of Zhoucun District in Zibo. Built in 1911.

## 烟台·烟台联合教堂旧址

位于烟台市芝罘区向阳街道办事处历新路街7号。建于1875年

Located at No.7 Lixin Road of the Xiangyang Sub-district Office of Zhifu District in Yantai. Built in 1875.

# 烟台·美国海军基督教青年会旧址

位于烟台市芝罘区向阳街道办事处海关街 26 号。建于 1921 年  Located at No.26 Haiguan Road of the Xiangyang Sub-district Office of Zhifu District in Yantai. Built in 1921.

教堂·庙观

烟台·卍字会旧址

位于烟台市芝罘区东山街道大马路十字街 51 号。建于 1925 年

Located at No.51 Shizi Road of the Damalu Residents' Committee of the Dongshan Sub-district Office of Zhifu District in Yantai. Built in 1925.

## 泰安·碧霞灵应宫

位于泰安市泰山区水帘洞坊北 30 米处。始建年代不详

Located around 30 meters to the north of Shuiliandong Arch Taishan District in Taian. When it was built is unknown.

## 泰安·无极庙

位于泰安市泰山区泰山扇子崖登山盘道起始处、竹林寺西 200 米处。建于民国时期

Located about 200 meter to the west of Zhulin Temple and at the beginning of the winding path for climbing the mount of the Shanzi Cliff of Mount Tai of Taishan District in Taian. Built during the period of Public of China.

## 泰安·小中泉天主堂

位于泰安市肥城市湖屯镇小中泉街居委会中心街北侧。建于1910年

Located to the north of the Center Street of Xiaozhongquan Residents' Committee of Hutun Town of Feicheng In Taian. Built in 1910.

## 泰安·新胜隆庄耶稣帝王堂

位于泰安市肥城市石横镇新胜居委会。建于 1840 年

Located in the Xinsheng Residents' Committee of Heng Town of Feicheng In Taian. Built in 1840.

## 泰安·玉都观

位于泰安市肥城市新城街道办事处孙家小庄村北长山街南首。建于1936年

Located at the south end of Beichangshan Road in Sunjiaxiazhuang Village of the Xincheng Sub-district Office of Feicheng In Taian. Built in 1936.

# 威海·宽仁院旧址

位于威海市环翠区竹岛街道办事处观海居委会海滨中路92号。建于1934～1937年

Located at No.92 Haibinzhong Road of the Guanhai Residents' Committee of the Zhudao Sub-district Office of Huancui District in Weihai. Built between 1934 and 1937.

## 滨州·高庙李天主堂

位于滨州市博兴县庞家镇高庙李村内。建于1935年

Located in GaolimiaoVillage of Pangjia Town of Boxing County in Binzhou. Built in 1935.

## 滨州·王家店子天主堂

位于滨州市惠民县麻店镇王家店子村内。建于民国时期

Located in Wangjiadianzi Village of Madian Town of Huimin County in Binzhou. Built during the period of Public of China.

## 滨州·小刘天主堂

位于滨州市阳信县阳信镇小刘村。建于1916年

Located in Xiaoliu Village of Yangxin Town of Yangxin County in Binzhou. Built in 1916.

图书在版编目(CIP)数据

百年掠影：山东近代建筑集萃 / 李华文主编. —济南：济南出版社，2014.1
 ISBN 978-7-5488-1190-9

Ⅰ.①百… Ⅱ.①李… Ⅲ.①建筑史-山东省-近代-图集 Ⅳ.① TU-092.5

中国版本图书馆CIP数据核字（2014）第005810号

| 书　名 | 百年掠影——山东近代建筑集萃 |
|---|---|
| 主　编 | 李华文 |
| 副主编 | 周玉山　马月华 |

责任编辑·装帧设计 / 戴梅海

翻译 / 崔振河　校译 / 崔　灿

封面题字 / 孙　猛

| 出　版 | 济南出版社 |
|---|---|
| 印　刷 | 济南黄氏印务有限公司 |
| 规　格 | 大12开（286×270毫米） |
| 印　张 | 15 |
| 版　次 | 2014年1月第1版 |
| 印　次 | 2014年1月第1次印刷 |
| 书　号 | ISBN 978-7-5488-1190-9 |
| 定　价 | 360.00元 |